Looking at Minibeasts

Flies and Mosquitoes

Sally Morgan

Belitha Press

Contents

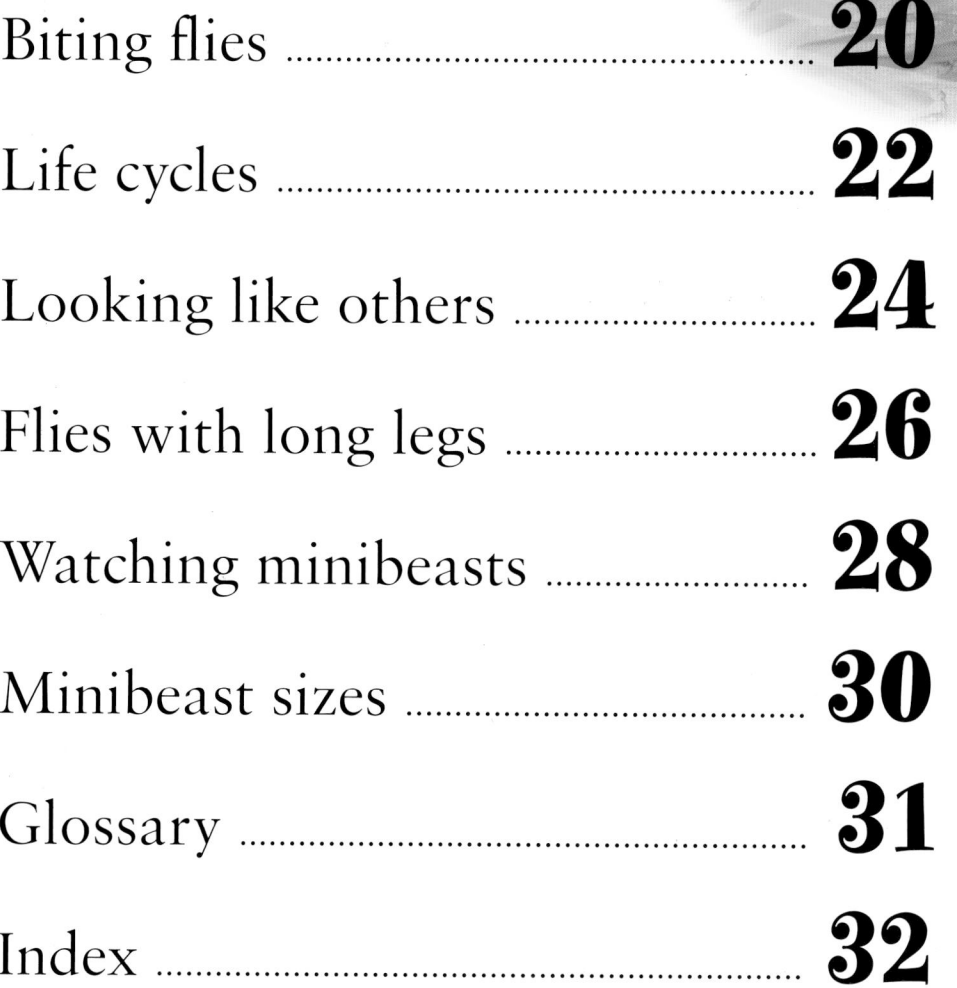

Words in **bold** are
explained in the
glossary on page 31.

What are flies and mosquitoes?

Flies and mosquitoes are **insects**. Their bodies are divided into three parts. They have a head with a pair of eyes and **antennae**, a thorax or chest with six legs and a back part called an **abdomen**. These insects have one pair of wings which are attached to the thorax.

Mosquitoes have six long legs attached to their thorax.

A mosquito is a biting insect. Its mouthparts are sharp enough to pierce skin.

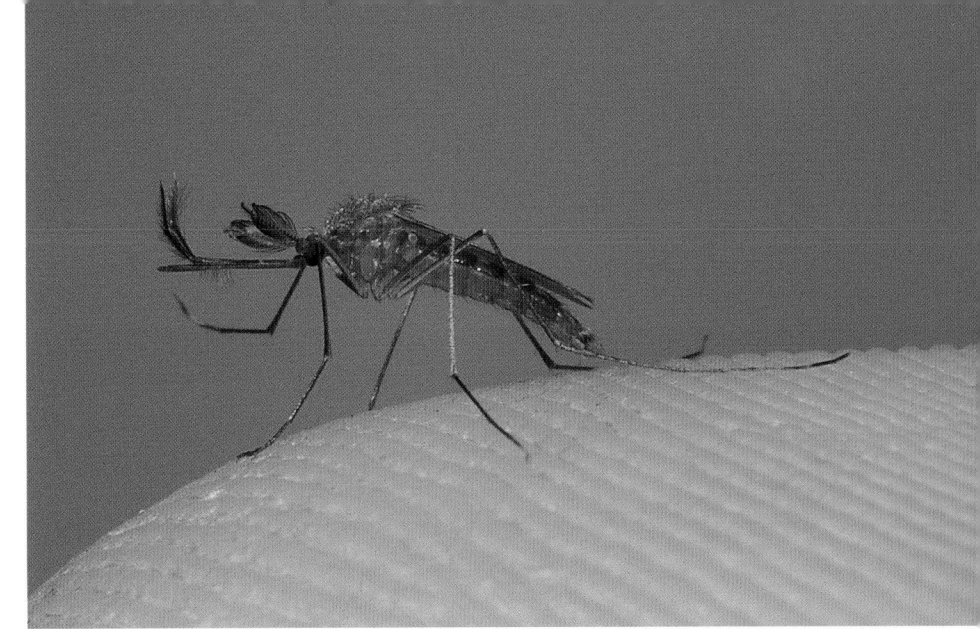

Large flies, such as this housefly, have hairy bodies. They have a large pair of eyes and a short pair of antennae.

The fly family

There are more than 90 000 different types of fly. They range in size from large horseflies to tiny midges. They are found all over the world, from very hot to very cold countries.

This is a giant fly. It is about 2 centimetres long and is the largest type of fly in the world.

Many flies are found around flowers, feeding on nectar.

Most insects have two pairs of wings, but mosquitoes and flies only have one pair. The second pair has been replaced by two knobbed stalks called **halteres**. An insect such as a butterfly has 'fly' in its name, but is not a true fly because it has four wings.

This long-legged fly is called a mimegralla. It is a true fly because it only has two wings.

Seeing and feeling

A fly has two large compound eyes. If you look very carefully at the eye of a fly you may be able to see that it is made up of many mini-eyes. A compound eye sees a picture that is made up of lots of tiny images. This means that flies cannot see things in detail, but they can see movement.

This close up shows how each eye is made up of mini-eyes arranged in rows.

The eyes of this fly are on the end of long stalks. They look like a pair of antennae.

Flies use their antennae, or feelers, to find food. Their antennae are short and hairy. They taste their food using their feet. Their feet are covered in taste buds – just like our tongues.

This close up of a fly's head shows its antennae, eyes and mouthparts.

Thin wings

The wings of a fly are thin and usually transparent. This means you can see through them. Flies can move their wings up and down very quickly. Some of the smallest flies beat their wings thousands of times each second.

A crane fly has long, delicate wings which are easily damaged.

10

This fly's long legs
hang down as it
flies around.

Hover flies can hover
in front of a flower
without moving.

The wings beat so fast that they make a buzz or whine. People know that mosquitoes are about at night because they can hear a whining sound as they fly around the room.

Sucking and biting

A fly has a mouth that can suck up food. It is called a **proboscis** and it looks like a tube with two suckers at the end. When a fly lands on food, it pours out the contents of its stomach. This turns the food into

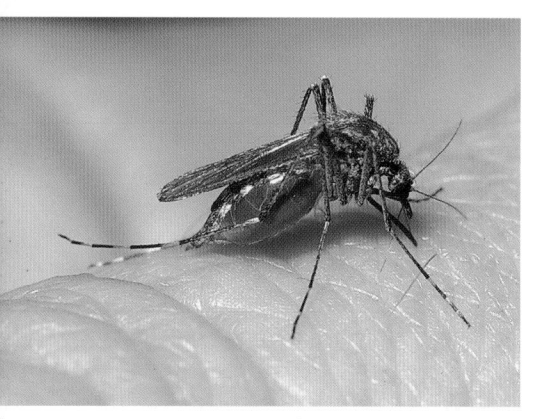

A mosquito pierces the skin with its mouthparts, then sucks up the blood.

a liquid which the fly can suck up. Mosquitoes feed on blood. They have long, sharp mouthparts, which they use to pierce skin.

Hover flies suck up the sugary
nectar found in flowers.

This empid fly is feeding
on a moth it has caught.

Houseflies

Many flies are found in and around the home. They are attracted to the food in the kitchen and rubbish bin. There are more flies around during the warm, summer months. Houseflies often have brightly-coloured bodies. Bluebottles and greenbottles are types of housefly.

On sunny days, the greenbottle can be seen buzzing around rubbish bins.

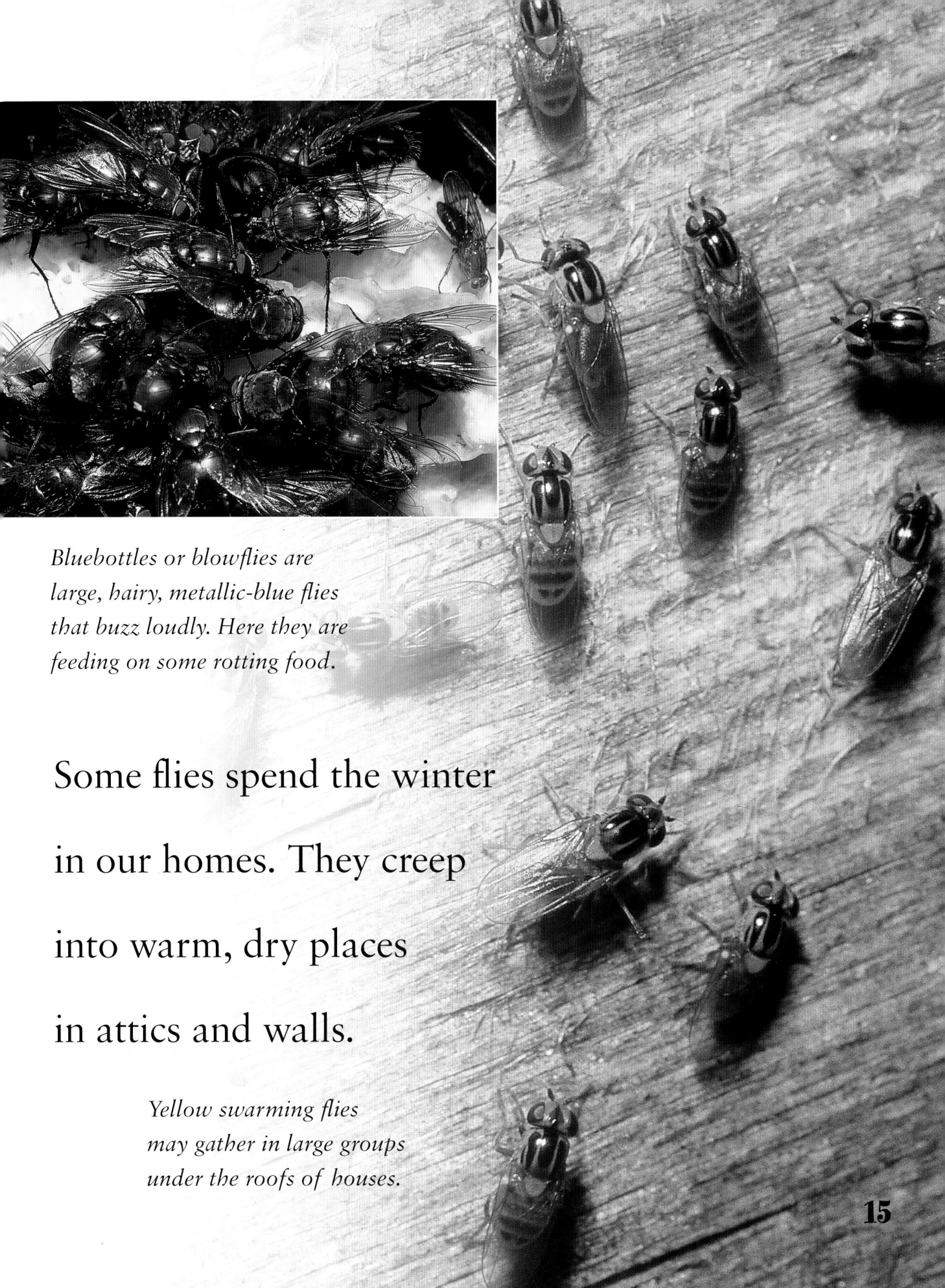

Bluebottles or blowflies are large, hairy, metallic-blue flies that buzz loudly. Here they are feeding on some rotting food.

Some flies spend the winter in our homes. They creep into warm, dry places in attics and walls.

Yellow swarming flies may gather in large groups under the roofs of houses.

Spreading disease

Houseflies are **pests** because they carry **disease**. A housefly has a hairy body, and when it walks over dirty surfaces or animal droppings, its hairs pick up dirt and germs. A fly constantly combs its hairs, dropping the dirt and germs on to food. This can spread diseases.

This fly is combing itself with its back legs. As it combs, germs and dirt will fall off its body.

Many flies, like these yellow dung flies, are attracted to dung. The flies may then land on food in kitchens.

People try to keep their kitchens free of flies by hanging up sticky fly papers which trap the flies. Or they use fly sprays which contain chemicals. The chemicals kill the flies within a few seconds.

This fly is sucking up sugar that it has found on a kitchen counter.

17

Maggots

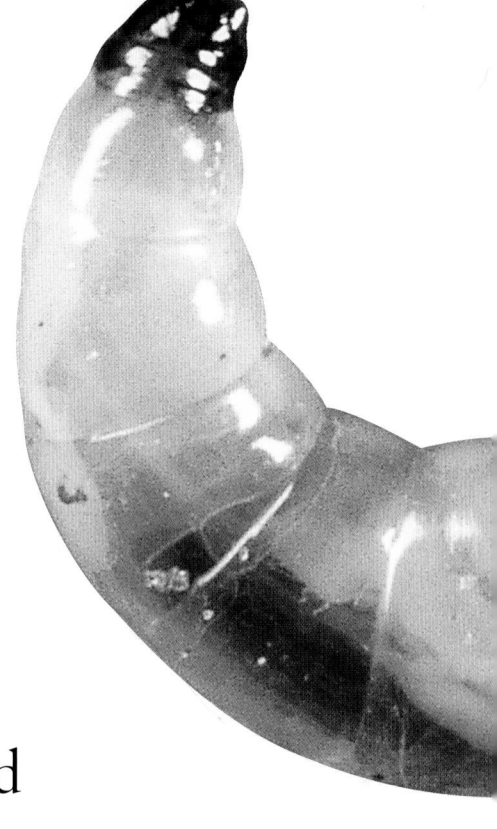

Flies help to break down dead bodies. The smell of a rotting animal body attracts many flies. The flies crawl over the body and lay their eggs. Each egg hatches into a soft-bodied **larva** called a **maggot**.

A fly's eggs hatch within a few days. Soon there is a wriggling mass of maggots.

18

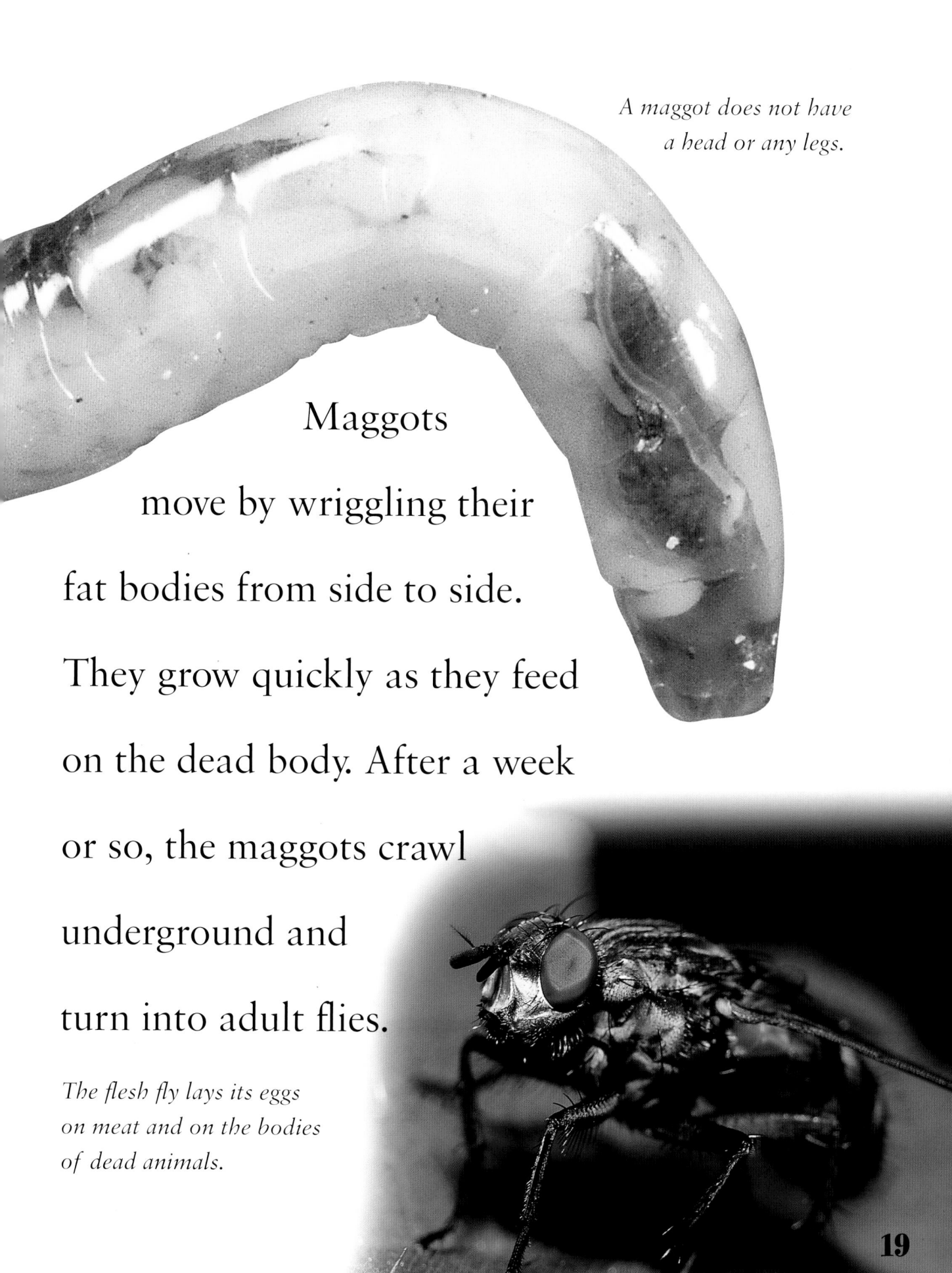

A maggot does not have a head or any legs.

Maggots move by wriggling their fat bodies from side to side. They grow quickly as they feed on the dead body. After a week or so, the maggots crawl underground and turn into adult flies.

The flesh fly lays its eggs on meat and on the bodies of dead animals.

Biting flies

Many flies are **carnivores**. This means that they feed on other animals. Robber flies and dung flies chase after smaller flies. They kill their **prey** and then suck out their body juices.

A yellow-bodied robber fly waits for a smaller insect to fly past. It grabs its prey with its strong legs and kills it.

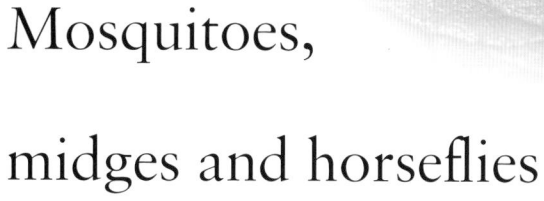

The mouthparts of this cleg fly work like a syringe to suck up the blood.

Mosquitoes, midges and horseflies bite people. They all need a meal of blood before they can lay their eggs.

In some countries, female mosquitoes carry a disease called malaria. If an **infected** mosquito bites a person, it can pass on the disease.

The horsefly is found around horses and cattle. It can give you a painful bite.

Life cycles

Mosquitoes, midges and gnats live near water. These flies lay their eggs in small pools or ditches where the water is still and not moving. The eggs float on the surface of the water.

A female gnat lays her eggs in water. The eggs form small rafts which float on the surface of the water.

The eggs hatch into larvae which live in the water. These larvae can be seen hanging at the surface of the water where they can breathe air. Each larva turns into a **pupa**. A few days later, the pupal case splits open and a new adult comes out.

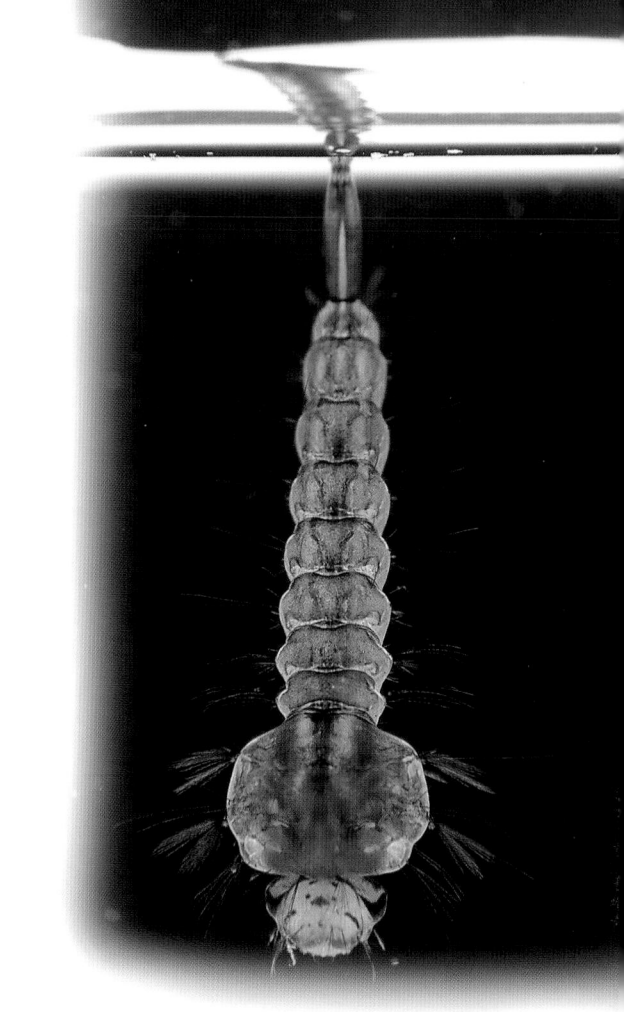

After it hatches, the larva (above) comes to the surface to breathe.

A new adult fly dries it wings before it flies away.

Looking like others

Many bees and wasps have brightly-coloured bodies. These are called warning colours because they warn other animals that they can sting. Some flies are the same colours as bees or wasps. This means that birds and other small animals that eat flies will leave them alone.

This yellow and black hover fly looks like a wasp. Birds avoid it because they think it can sting.

24

These drone flies look like honeybees. They visit flowers at the same time as bees.

There are hover flies that look just like wasps and honeybees. The bee fly even has a hairy body like a bee. You will see these flies when there are plenty of bees and wasps around.

The bee fly not only looks like a bee, it flies from flower to flower and sucks up nectar like a bee too.

Flies with long legs

The fly with the longest legs is the crane fly, or daddy-longlegs. It has a long body too. Crane flies appear in the early evening. They are attracted by lights in homes. This large fly may look scary, but it is harmless.

A crane fly's long legs break off easily so that it can escape if it is caught by a predator.

The larva of a crane fly is called a leatherjacket. It lives in the soil and eats the roots of plants.

Some other long-legged flies have heavy bodies, so they fly slowly. As they fly through the air, their long legs hang down. Midges are small flies, and they have long legs too.

On a summer's evening, large groups of midges and gnats appear near water. Gnats make a buzzing sound, but midges are silent.

Watching minibeasts

Summer is the best time of year to look out for flies and mosquitoes.

Angling shops sometimes sell maggots as fish bait. You can watch how the maggots wriggle around. Try this simple test to see if maggots like the light. Place a large piece of paper on a table. Position a desk light so that it lights up half of the paper. Put a few maggots in the middle of the paper and see if they move towards or away from the light. Remember not to keep the maggots for too long or they will turn into pupae and become flies.

Watch how the maggots behave when they are placed under a light.

Gnats lay their eggs in water.

If you leave a small container of water outside during summer, it will attract gnats. The gnats will lay their eggs on the surface of the water. Within a few days you may be able to spot small animals swimming around in the water. These are probably gnat larvae.

Flies can carry germs that cause disease, so people do not like to have them in their homes. Some houses have screens or nets across the doors and windows to keep flies out. Fly sprays and sticky fly papers help to keep a home free of flies.

People cover food with cloths or nets to stop flies touching the food.

Minibeast sizes

Flies and mosquitoes are many different sizes.
The pictures in this book do not show them
at their actual size. Below you can see how
big some of them are in real life.

Hover fly
13 millimetres long

Maggot
10 millimetres long
(when nearing full
size)

Giant fly
20 millimetres long

Cleg fly
10 millimetres long

Crane fly
19 millimetres long

Mosquito
5 millimetres long

Leatherjacket
25 millimetres long

Glossary

abdomen The rear or back part of an insect's body.

antennae The feelers on an insect's head.

carnivores Animals that eat other animals.

disease An illness, usually caused by germs.

halteres Parts of a fly that look like knobbed stalks. They help a fly to balance while it is flying.

infected Carrying germs.

insects Animals with six legs and three parts to their bodies.

larva (plural: larvae) The young form of an insect. It looks different to an adult.

maggot The larva of a fly.

pests Animals that damage crops, homes, or carry germs and disease.

prey Animals that are killed by other animals for food.

proboscis (say: *pro-boss-iss*) A long mouthpart for piercing or sucking food.

pupa A hard case made from the skin of a larva. The larva turns into an adult inside the pupa.

Index

Editors: Claire Edwards, Sue Barraclough
Designer: John Jamieson
Picture researcher: Sally Morgan
Educational consultant: Emma Harvey

First published in the UK in 2001 by

Belitha Press Limited
London House, Great Eastern Wharf,
Parkgate Road, London SW11 4NQ

Copyright © Belitha Press Limited 2001
Text copyright © Sally Morgan 2001
Illustrations by Woody

ISBN 1 84138 351 1

Printed in Hong Kong

British Library Cataloguing in Publication Data
for this book is available from the British Library.

10 9 8 7 6 5 4 3 2 1

Picture acknowledgements:
Greenwood/Ecoscene: 24. Chinch Gryniewicz/Ecoscene:
25b. Wayne Lawler/Ecoscene: 26-27. Papilio: front &
back cover tcr & c, 5t, 5b, 18, 19b, 21t, 21b, 23t, 23b,
30clb. K.G. Preston-Mafham/Premaphotos: front cover
tr & cl & cr, 2, 3t, 4, 6, 7b, 9t, 10, 11tl, 12, 13t, 13b,
15t, 15r, 16, 17t, 20b, 22, 25t, 26b, 30cl, 30cb, 30crb.
M. Preston-Mafham/Premaphotos: 1, 14. R.A. Preston-
Mafham/Premaphotos: front & back cover tl & tcl, 7t,
11tr, 17b, 30c. Kjell Sanders/Ecoscene: 3b, 8, 9b.
Robin Williams/Ecoscene: 19t, 27t, 30cr, 30b.